Thomas Edison

For Alex

Oxford University Press, 198 Madison Avenue, New York, New York 10016

Oxford   New York

Athens   Auckland   Bangkok   Bogotá   Bombay
Buenos Aires   Calcutta   Cape Town   Dar es Salaam
Delhi   Florence   Hong Kong   Istanbul   Karachi
Kuala Lumpur   Madras   Madrid   Melbourne
Mexico City   Nairobi   Paris   Singapore
Taipei   Tokyo   Toronto   Warsaw

and associated companies in
Berlin   Ibadan

Oxford is a trademark of Oxford University Press

Text © Haydn Middleton 1997
Illustrations © Oxford University Press 1997

Originally published by Oxford University Press UK in 1997.

Library of Congress Cataloging-in-Publication Data
Middleton, Haydn.
    Thomas Edison / Haydn Middleton : illustrated by Tony Morris.
        p.   cm. – (What's their story?)
    Includes index.
        1. Edison, Thomas A. (Thomas Alva), 1847 – 1931 –Juvenile
literature.   2. Electric engineers – United States –Biography –
Juvenile literature.   3. Inventors – United States – Biography –
Juvenile literature.   [1. Edison, Thomas A. (Thomas Alva),
1847–1931.   2. Inventors.]   I. Morris, Tony, ill.   II. Title.
III. Series.
TK140.E3M645   1997
621 .3'092--dc21
    [B]   97–27373
                                                                           CIP
                                                                           AC

ISBN 0-19-521401-3

1 3 5 7 9 10 8 6 4 2

Printed in Dubai by Oriental Press

Thomas Edison

THE WIZARD INVENTOR

HAYDN MIDDLETON
Illustrated by Tony Morris

OXFORD UNIVERSITY PRESS

A hundred and fifty years ago in Milan, Ohio, Samuel and Nancy Edison had a baby boy. They named him Thomas. They worried about him when he was small, because he was often ill. He was so sick that he could not go to school until he was eight. Then he started to go deaf. But Tom Edison was going to live to the grand old age of 84—and when he died, his name would be known all over the world.

Tom grew up close to the Canadian border. Life was very different then. Native American tribes still roamed America's Great Plains. Black slaves worked in the cotton fields. Scientists knew about electricity, but no one had figured out how to use it in people's homes. So there were no televisions, no telephones, no stereos, not even any electric lights.

Young Tom looked at this world, and he began to ask questions.

"What makes birds fly?" "How does fire work?" "Why is the sky blue?" Tom was always asking questions. This annoyed his teachers. They wished he would just sit quietly like the other children. Tom's mother took him out of school and tried to teach him herself. When she could not answer his questions, Tom began to make his own investigations.

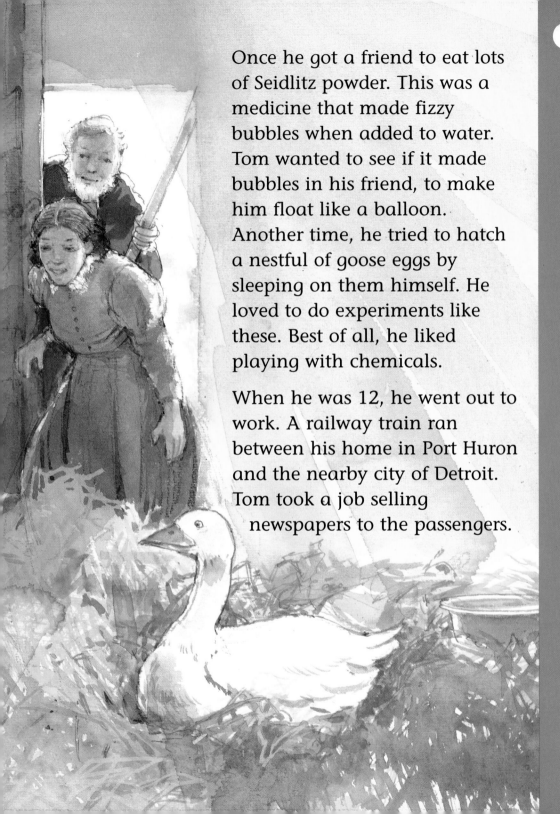

Once he got a friend to eat lots of Seidlitz powder. This was a medicine that made fizzy bubbles when added to water. Tom wanted to see if it made bubbles in his friend, to make him float like a balloon. Another time, he tried to hatch a nestful of goose eggs by sleeping on them himself. He loved to do experiments like these. Best of all, he liked playing with chemicals.

When he was 12, he went out to work. A railway train ran between his home in Port Huron and the nearby city of Detroit. Tom took a job selling newspapers to the passengers.

Tom's idea of a good time was to work from dawn to dusk. He was more than just a newspaper boy on his train trips to and from Detroit. He also sold candy and drinks, and vegetables from his parents' garden. For a while he even wrote and printed his own little paper, full of local news and gossip. He would spend his lunch times at the Detroit Free Library. His aim was to read every single book there—even the ones he did not really understand!

If he ever had a free moment on the train, he conducted experiments with his chemicals. The guard had given him permission to set up his equipment in a quiet corner. Tom labeled each of his bottles POISON to stop anyone else from playing around with them.

Unfortunately, one of his experiments went wrong. There was a bang and the smoking car caught fire. The guard was furious. He threw out all Tom's equipment. But not all the railway workers were so unhelpful.

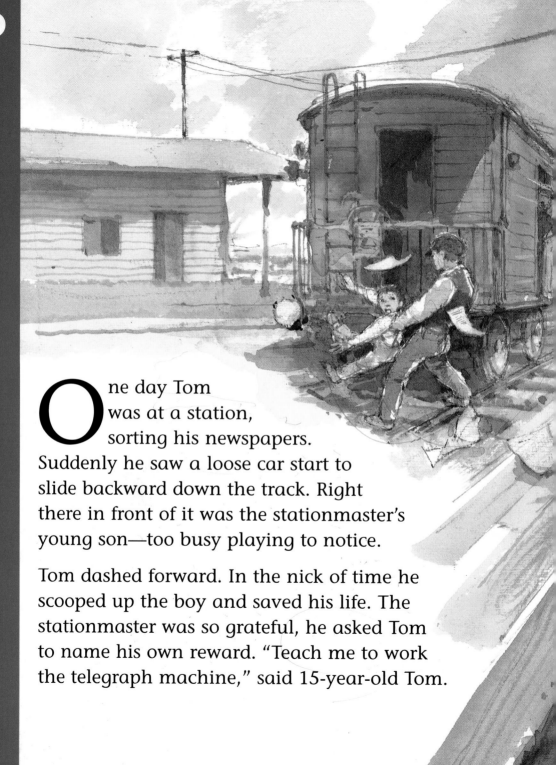

One day Tom was at a station, sorting his newspapers. Suddenly he saw a loose car start to slide backward down the track. Right there in front of it was the stationmaster's young son—too busy playing to notice.

Tom dashed forward. In the nick of time he scooped up the boy and saved his life. The stationmaster was so grateful, he asked Tom to name his own reward. "Teach me to work the telegraph machine," said 15-year-old Tom.

In those days the telegraph machine was the quickest way of sending a message over a long distance. You could tap out a message at one end of an electric wire and someone miles away at the other end would receive it. The message was sent in Morse code—an alphabet of dots and dashes. Telegraph wires crisscrossed the entire United States. Now Tom could find out exactly how the messages were sent.

By the age of 16, Tom had learned how to be a telegraph operator. He was still deaf, but he could hear the clicks on the line well enough. So he decided to stop working on the railway. For the next six years he traveled from one city telegraph office to another, earning his living by sending and receiving messages.

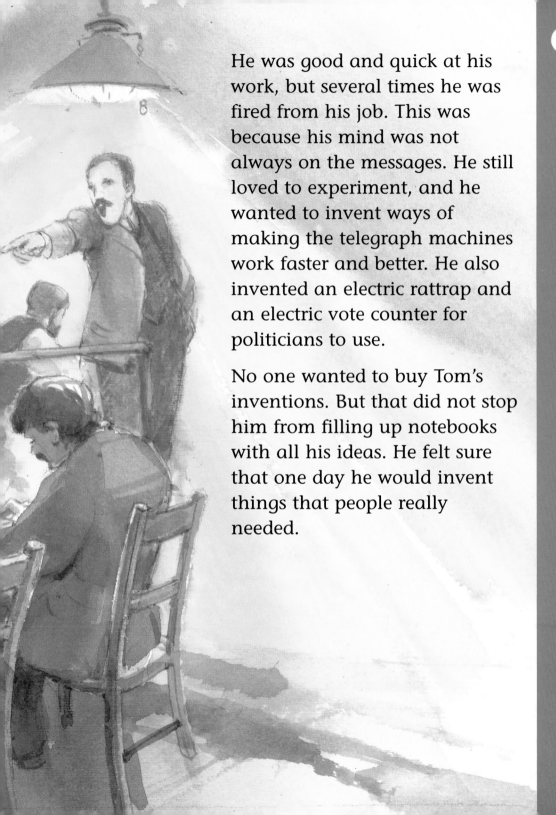

He was good and quick at his work, but several times he was fired from his job. This was because his mind was not always on the messages. He still loved to experiment, and he wanted to invent ways of making the telegraph machines work faster and better. He also invented an electric rattrap and an electric vote counter for politicians to use.

No one wanted to buy Tom's inventions. But that did not stop him from filling up notebooks with all his ideas. He felt sure that one day he would invent things that people really needed.

Tom's travels took him to New York. He was getting bored with tramping around the country as a telegraph operator. What he really wanted to be was an inventor and machine maker. So he showed the Western Union Telegraph Company some new ways to make their telegraph machines work better.

This time he was lucky. Western Union paid him well for all his ideas—with a check for $40,000. Tom used the money to set up a workshop of his own in nearby Newark. Then, for five years, he worked harder than ever, mainly making improvements to the telegraph machine.

He hired engineers to work for him. They called him the Old Man, even though he was still in his twenties. He could be a fierce boss, but fun too. Whenever an improvement worked, he did a little Zulu war dance to celebrate! In 1871, when he was 24, Tom had something else to celebrate. He got married.

Tom's new wife was Mary Stillwell. She was only 16, and she thought Tom was a wonderful, clever, funny man. But she was also a little afraid of him, and he was always so busy! Even on their wedding day he worked until midnight. Sometimes she did not see him for days. The main thing in Tom's life was his work. It always had been, and it always would be.

Tom and Mary had three children: Marion, Thomas, and William. Tom nicknamed the first two Dot and Dash, like the Morse code. He often let the children play near him while he worked. This was no longer in Newark. That old workshop had become too small. So Tom had paid his father to come from Port Huron and build him a big new workplace at Menlo Park in New Jersey. He called it his "inventions factory."

Tom began to work at Menlo Park in 1876. His next few inventions were going to make him famous. People started calling him the Wizard of Menlo Park. They thought his new sound and light inventions were like wonderful magical spells. But even the best wizard needs helpers. Tom knew this. He paid several brilliant men to come and work with him at his inventions factory.

One of these men was John Kreusi, from Switzerland. Another was Charles Batchelor, an engineer born in England. As time went by, more men joined the team. Often they had to work together right through the night. But they had a lot of laughs, too. Tom kept everyone smiling with his jokes. He even put an organ in the laboratory, for music during mealtimes. And what did Tom want from them? "A minor invention every 10 days and an important one every 6 months." The first important one came very soon.

Alexander Graham Bell had just made the first "speaking telegraph," or telephone. It was a wonderful thing, but the voices sounded faint, and they carried for only a few miles. Western Union wanted someone to test it and make it work clearly over long distances. So they asked the Wizard of Menlo Park.

Tom and his team conducted more than 2,000 experiments with the telephone. Tom's hearing was now so bad that he had to use his teeth to listen with. He attached a magnet to the phone, bit on it—and the sound waves passed through his jaw to the inner parts of his ears, which still worked!

At last he made the breakthrough. He invented a small carbon transmitter that made all the difference. Now even he could hear someone speaking on the phone. Western Union gladly bought his invention for $100,000, and all over the world the Age of the Telephone began. Tom probably did his little Zulu dance to celebrate. But he already had another brilliant idea.

Tom wanted to make a "phonograph," or sound writer: a machine that could record and play back the human voice. Even Tom's team thought this was impossible.

But with a sharp-tipped carbon transmitter, Tom recorded his voice on to a cylinder wrapped in tinfoil. When he passed the cylinder under the tip again, the words were played back. The first words his team heard from the machine were "Mary had a little lamb...." They thought it was a trick. Surely someone was hiding in the room and echoing what Tom had said!

It took 10 more years to make phonographs good enough to sell. Before then, the President invited Tom to the White House to give him a personal performance. "I've made a good many machines," Tom said, "but this is my baby, and I expect it to grow up to be a big fellow." He was right. Today's huge recording industry began with his phonograph. But his last great invention would have an even bigger effect.

A hundred and twenty years ago, the world was a darker place. Gas lights or powerful electric "arc" lamps burned on some streets. But after the sun went down, most people lit their homes with weak candles or smoky, smelly oil lamps. Both could set houses on fire if someone knocked them over.

Like many other scientists, Tom dreamed of putting glowing electric lights into even the poorest people's houses. Now he boldly declared that he would make this dream come true.

First he had to make a lightbulb that would glow for hours when switched on. Thousands of experiments later, he had one. He used carbonized, or sooty, cotton as the bulb's filament. This was the thread that heated up and glowed brightly. Later he used carbonized bamboo. He lit all of Menlo Park with these bulbs. But that was just the beginning. His plan was to light up all of New York City.

It was 3 P.M. on Monday, September 5, 1882. Nearly three years had passed since Tom had invented his lightbulb. Now he stood in his great new electric power plant on Pearl Street in New York.

He was very excited but very nervous, too. The moment had come to open the power plant. If everything went according to plan, electricity would flow from it through miles of underground cables into the homes of just 85 paying customers. Their houses had been specially wired and fitted with lightbulbs. Would the bulbs light? If they did, millions of people would want their homes to be connected to power plants in the same way.

Tom nodded at the chief electrician. "Pull the switch," he said. A moment later, the lights went on in 85 different places. As everyone cheered, the Wizard Inventor could relax. Thanks mainly to him, the Age of Electric Light and Power had begun. But for Tom, sadness lay ahead.

Tom moved his family to New York while he worked on the power plant. Two years after it opened, his wife, Mary, died of typhoid. Tom and his children felt lost. They thought that no one could take her place.

Then Tom met a beautiful young woman named Mina Miller. Although he was very deaf, they talked by tapping Morse code into each other's hands. One day in a busy railway car Tom tapped "Will you marry me?" into Mina's hand. "Yes," she tapped back. And no one else in the car knew what had happened.

Tom, Mina, and the children moved to a big new home and inventions factory in West Orange, New Jersey. Three more children were born: Charles, Theodore, and Madeleine. For years, Tom kept working as hard as ever. He had some small successes, but by now his great inventing days were over.

"Tom invents all the time," Mina Edison said, "even in his dreams." He and his team had made no fewer than 1,093 inventions. Some were improvements on the work of other people. Sometimes people improved on Tom's inventions, too—like his kinetoscope, an early machine for showing movies.

Tom was a confident person. "Anything, everything, is possible," he said. People called him a wizard, but he always had a purpose in mind for his magic. That was why he was once voted "America's Most Useful Man."

Tom died in 1931, at the age of 84. He had lit up the world—there had to be a special way of saying good-bye to him. Three nights later, at 10 o'clock, people all over the United States switched off their electric lights for one minute. Even the Statue of Liberty's torch went dark. Then the lights returned.

The great inventor was dead. His inventions would live on.

Important dates in Thomas Edison's life

Index